T0146384

John Fowler, Benjamin Baker Forth Bridge

Texts
Iain Boyd Whyte
Angus J. Macdonald

Photographs
Colin Baxter

Edition Axel Menges

Editor: Axel Menges

© 2012 Edition Axel Menges, Stuttgart/London
ISBN 978-3-930698-18-9

Second, revised edition

Reproductions: Bild und Text GmbH Baun, Fellbach
Printing and binding: Graspo CZ, a. s., Zlín, Czech
Republic

Design: Axel Menges

Contents

Iain Boyd Whyte
A sensation of immense power

The engineers, with their gigantic works, sweep everything before them in this Victorian era.
(Benjamin Baker)

The railway bridge across the River Forth was opened in 1890. In the same year its first and best biography was published. It was written by Wilhelm Westhofen, who had worked as assistant engineer on the bridge, with special responsibilities for the foundations and masonry piers. With admirable thoroughness Westhofen's account covered everything remotely connected with the bridge, ranging from the pilgrimages made across the river from Edinburgh to Dunfermline in the twelfth century by Margaret, King Malcolm Canmore's second wife, to the expansion joints in the steelwork. Lengthy biographies of the main protagonists in the tale were also included. Westhofen's text was complemented by the glass-plate negatives made by Evelyn Carey, who had been commissioned to make a complete photographic record of the bridge's construction. Taken together, the text and photographs give the most complete description of any of the great works of nineteenth-century engineering. Given the virtual impossibility of improving on these accounts, subsequent histories of the bridge have leaned heavily on Westhofen and Carey, with the same sequence, text and illustrations appearing again and again, with varying degrees of acknowledgement to the original source. The Westhofen/Carey recycling industry reached its zenith in 1990, when the centenary celebrations for the bridge stimulated a plethora of publications that were indebted to these nineteenth-century antecedents.[1]

Content in the knowledge that the history of the bridge has already been described in detail by Westhofen/Carey and their followers, this text concentrates on the question: why do we respond to it as we do?

Early in November 1880 the body of the bridge engineer Sir Thomas Bouch was brought to Edinburgh for burial in the Dean Graveyard. As he was laid to rest, the east side of the graveyard was guarded by Thomas Telford's great Dean Bridge, built around 1800 on four tall masonry arches that rise 100 feet above the Water of Leith, while to the north the funeral party would have seen the silvery light of the Firth of Forth. Seven miles upstream from Edinburgh at South Queensferry, abandoned workshops marked Sir Thomas's last, doomed project, for a suspension bridge across the Forth. Together with his bridge over the River Tay, which had been opened for passenger traffic in March 1878, the two bridges were designed to vault the two wide estuaries that cut deeply into the North Sea coast of Scotland, driving the east-coast railway traffic into the centre of the country. With these two major natural obstacles overcome, an unbroken railway link could run up the eastern side of Britain from Dover to Aberdeen. In the local context, the bridges over the Forth and Tay would banish the difficult and sometimes dangerous sequence of three separate trains and two ferries necessary in 1870 to take goods and passengers from Edinburgh to Dundee, a direct distance of only 40 miles.

Bouch's world, however, tumbled down around him when the Tay Bridge was blown down during a gale just after 7 pm on Sunday 28 December 1879, taking a passenger train with it. There had been disasters with railway bridges before, the price of rapacious entrepreneurialism and an infant technology. The weight of the trains crossing a suspension bridge built by Captain Samuel Brown in 1830 to carry the Stockton and Darlington Railway across the River Tees caused a wave effect in the deck that racked the bridge to pieces in a few years. More spectacularly, the collapse in May 1847 of the Dee Bridge, a trussed-girder design by Robert Stephenson, led to such consternation about the safety of railway bridges that a parliamentary commission was set up to investigate the new iron technology.[2] Yet none of these earlier disasters caught the public imagination more vigorously, both in Britain and throughout the world, than the collapse of the Tay Bridge, with the loss of 75 lives.

While the popular press fulminated and the technical journals offered detailed analyses of the causes of the collapse, the poets set to with varying degrees of sophistication. Shortly after its official opening in March 1878, Dundee's most celebrated and least talented poet, William McGonagall, had lauded the new bridge as:

> The greatest wonder of the day,
> A great beautification to the River Tay.

Now, less than two years later, the homespun »Poet and Tragedian« was saddened to record:

> Beautiful Railway Bridge of the Silv'ry Tay!
> Alas! I am very sorry to say
> That ninety lives have been taken away
> On the last Sabbath day of 1879
> Which will be remember'd for a very long time.[3]

More tellingly, perhaps, the great Prussian novelist and poet Theodor Fontane penned his poem »Die Brück' am Tay« on 6 January 1880. While in McGonagall's account »Boreas blew a terrific gale«, Fontane has the three witches from Macbeth conspiring to meet at the central pier of the bridge at 7 o'clock to sweep both the structure and the scheduled train into the briny, chanting: »Trash, trash / Is the product of human hand!« (»Tand, Tand / Ist das Gebilde von Menschenhand!«).[4]

The Report of the Court of Enquiry, which gave its verdict in June 1880, placed the prime responsibility not on Macbeth's witches, but on Sir Thomas Bouch: »We find«, stated the report, »that the bridge was badly designed, badly constructed and badly maintained. ... For these defects ... Sir Thomas Bouch is, in our opinion, mainly to blame.«[5] The resulting odium not only killed Bouch, but added to public concern about the dangers of railway travel, a subject that had already stimulated learned medical debate in the 1860s and 1870s. Indeed, the railway accident was seen as the ultimate condition of trauma, both physical and psychological. »It must be obvious«, wrote John Erichsen in his book *On Railway and Other Injuries of the Nervous System*, »that in no ordinary accident can the shock be so great as in those that occur on railways. The rapidity of the movement, the momentum of the person injured, the suddenness of its arrest, the help-

1. Eric de Maré, steam train crossing the Forth Bridge, 1952. (Scottish Record Office.)
2. Thomas Bouch, unbuilt project for the Forth Bridge. (Angus J. Macdonald.)

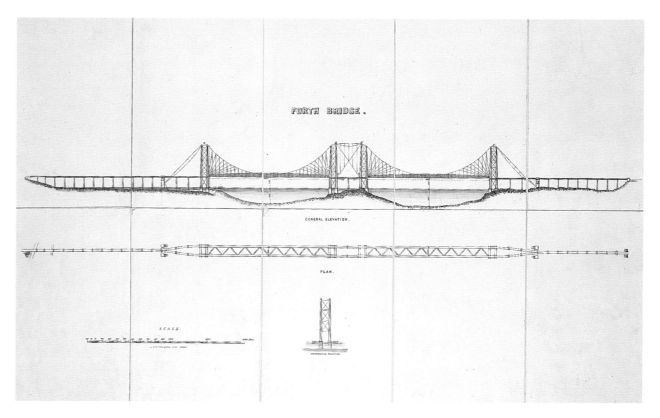

FORTH BRIDGE.

GENERAL ELEVATION.

PLAN.

SCALE.

lessness of the sufferers, and the natural perturbations of mind that must disturb the bravest, are all circumstances that of a necessity greatly increase the severity of the resulting injury to the nervous system.«[6] With the poets, the medics, and the popular press fixated on the horrors of the railway crash, it is not surprising to find in Lewis Carroll's *Through the Looking Glass*, written in 1872, Alice's dream of a railway journey rudely terminated by »a shrill scream from the engine«, and the news that the train was about to jump over a river. Alice, we are told »felt a little nervous at the idea of trains jumping at all«, but »in another moment she felt the carriage rise straight into the air«.[7] At that point she woke from her nightmare. In the real world of iron and steam, the consequences of railway accidents were infinitely more terrible, and it was in this climate of fear and distrust that Sir John Fowler and Benjamin Baker submitted their proposal for a new bridge to the directors of the Forth Bridge Railway Company on 30 September 1881.[8]

Previous proposals for a river crossing had favoured suspension bridges. Bouch's design envisaged a two-span bridge supported by four pylons, while an even earlier project put forward by James Anderson in 1818 had proposed a three-span structure. As Westhofen noted of Anderson's scheme: »To judge by the estimate, the designer can hardly have intended to put more than from 2,000 to 2,500 tons of iron into the bridge, and this quantity distributed over the length would have given the structure a very light and slender appearance, so light indeed that on a dull day it would hardly have been visible, and after a heavy gale probably no longer to be seen on a clear day either.«[9] In the aftermath of the Tay Bridge disaster a more substantial structure was demanded to reassure both the investors and the travelling public.

Exactly this was provided by the final version of the bridge by John Fowler and Benjamin Baker, which was adopted by the Forth Bridge Company in June 1881. Their design was evolved from a consideration of purely technical factors connected with functional requirements, the characteristics of the site, constructional considerations and the need to provide for an efficient use of structural material. All these parameters can be read in the configuration of the structure, which is remarkable for its simplicity and for the clarity with which its structural function was expressed. This clarity was accorded its own aesthetic status, and no architectural decoration or ornamentation of any kind was admitted to the design. In this sense, the bridge may be regarded as a work of pure engineering.

Precisely this lack of decorative embellishment appealed to the architect Alfred Waterhouse, who wrote to John Fowler: »One feature especially delights me – the absence of all ornament. Any architectural detail borrowed from any style would have been out of place in such a work. As it is, the bridge is a style unto itself: the simple directness of purpose with which it does its work is splendid and invests your vast monument with a kind of beauty of its own, differing though it does from all the beautiful things I have ever seen.«[10] Waterhouse's sentiments were also shared by the organisers of the official opening of the bridge. As *The Times* reported: »Fortunately no attempt was made to desecrate the bridge itself with flimsy adornments. It was allowed to stand out in simple and impressive grandeur.«[11]

The opening was conducted on 4 March 1890 by the Prince of Wales, surrounded by a distinguished company of railway barons, financiers and engineers, both British and continental. Gustave Eiffel, the great French engineer and designer of the eponymous tower, was also present to watch the Prince drive in the last rivet at the centre of the north span. The subseqent opening speech that was to be given in the stone arch at the southern end of the bridge became the victim of the high winds blowing on that day, and was shortened to a mere ten words: »Ladies and Gentlemen, I now declare the Forth Bridge open.« The stability of the bridge in the face of the storm was

3. Evelyn Carey, general view of the cantilevers in construction, 1887. (Scottish Record Office.)
4. Evelyn Carey, Queensferry pier from the river, 23 May 1887. (Scottish Record Office.)
5. Evelyn Carey, side elevation of one of the cantilevers, undated. (Scottish Record Office.)

the subject of many laudatory critiques of the new bridge, whose silent sub-text was the fate of the Tay Bridge ten years earlier. As the journal *Industries* reported: »With the wind whistling among the wind bracings, and ineffectually spending its fury on the top members of the bridge, more than a 100 feet above the viaduct, the bridge was opened in the middle of an appropriately severe trial. The perfectly calm security of the structure while so severe a gale was blowing must have impressed everyone who was present on this historic occasion.«[12] In the national press too, the plaudits were loud and enthusiastic. *The Times*, for example, devoted a long descriptive article to the bridge on the day of the official opening, concluding with some thoughts on its aesthetic qualities. »Few persons«, said *The Times*, »probably will be disposed to regard the Forth Bridge as a beautiful structure. It cannot compare in this respect with the Brooklyn Bridge, which is a graceful and well-proportioned erection. Nevertheless, if regard be had to conditions of strength and stability, to the adaption of means to ends, to the massiveness of the whole and the symmetry of the parts, it will be universally admitted that there are few human structures, if any, that surpass the Forth Bridge in impressiveness and stately grandeur.«[13] The notion that a purely utilitarian object could be the source of aesthetic delight or even be regarded as high art was a difficult one for the contemporary intellect. For throughout the nineteenth century it had been a basic premise of aesthetics that utility and functionality precluded admission to the pantheon of high art. Back in the 1790s, Immanuel Kant drew a firm distinction between free beauty (*pulchritudo vaga*) and dependent beauty (*pulchritudo adhaerens*), and consigned works of architecture to the latter and lesser category. As Kant's definition of the beautiful insists: »Beauty is the form of purposiveness in an object so far as this form is perceived in it without the concept of purpose.«[14] Hegel also supported this position in his *Aesthetics*, in which he maintained that art can serve no ulterior functions and must remain an end in itself. As the century of engineering progressed, however, the charms of a theoretical aesthetics that appeared to value decorative friezes more than the Parthenon were challenged by the designers.

As its name would suggest, the relationship between high art and the work of the engineer was a central concern of the National Association for the Advancement of Art and its Application to Industry. By chance, the annual meeting of this interest group was held in Edinburgh in 1889, with the almost completed Forth Bridge as its polemical coulisse. Probably inspired by the great structure that was taking shape at the edge of his native city, the eminent Scottish architect Rowand Anderson proposed a new beauty, that of the machine. In words that prefigured by 20 or 30 years the modernist polemics of Walter Gropius and Le Corbusier, Anderson hymned the beauty that derived from the purposeful power of the machine. »Who has looked down into the engine room of one of the great ocean steamers«, he asked, »and not felt the impression of an irresistible power that rests not day nor night? ... The designing of machinery, whether for peace or war, has now reached such a high standard of excellence in function, form, and expression that one is justified in saying that these things are entitled to rank as works of art as much as a painting, a piece of sculpture, or a building.«[15]

One speaker at the same meeting, however, vigorously rejected the idea of an art of the engineer, citing the Forth Bridge in evidence. This was William Morris, the leading light of the English Arts and Crafts movement, who claimed: »As for an iron architecture, there never was and never could be such. Every improvement in the art of engineering made the use of iron more ugly, until at last they had that supreme specimen of ugliness, the Forth Bridge.«[16] Morris's own preferences in the way of bridges can be adjudged from his novel *News from Nowhere*, published a year later in 1890. In this neo-medievalist utopia, Morris describes a bridge built in the year 2003 along fourteenth-century lines: »I had perhaps dreamed of such a bridge, but never seen such a one out of an illuminated manuscript; for not even the Ponte Vecchio at Florence came anywhere near it. It was of stone arches, splendidly solid, and as graceful as they were strong. ... Over the parapet showed quaint and fanciful little buildings, which I supposed to be booths or shops, beset with painted and gilded vanes and spirelets. ... In short, to me a wonder of a bridge.«[17] With his head full of vanes and spirelets, it is hardly sur-

prising that Morris had little enthusiasm for the steely magnificence of the Forth Bridge.

Morris's attack prompted a vigorous response from Benjamin Baker, delivered in the context of a lecture to the Edinburgh Literary Institute, in which he questioned if Morris had »the faintest knowledge of the duties which the great structure had to perform«. Sharpening his attack, Baker suggested: »Probably Mr. Morris would judge the beauty of a design from the same standpoint, whether it was for a bridge a mile long, or for a silver chimney ornament.« Baker went on to counter the criticism that the massive tubes forming the lower lateral members of the bridge should have formed a true arc, by insisting: »Critics must first study the work to be done by the piers, and by the superstructure, and also the materials employed, before they are capable of settling whether it is beautiful or ugly. »It would«, he added, »be a ludicrous error to suppose that Sir John Fowler and he had neglected to consider the design from the artistic point of view. They did so from the very first. An arched form was admittedly graceful, and they had approximated their bridge to that form as closely as they could without suggesting false constructions and shams. They made the compression members strong tubes, and the tension members light lattice work, so that to any intelligent eye the nature of the stresses and the sufficiency of the members of the structure to resist them were emphasized at all points. ... The object had been so to arrange the leading lines of the structure as to convey an idea of strength and stability. This, in such a structure, seemed to be at once the truest and highest art.«[18]

The debate over the Forth Bridge that had marked the Edinburgh meeting of the National Association for the Advancement of Art rumbled on at the next annual meeting, held in Birmingham in 1890. Indeed, in his presidential address, J. E. Hodgson identified in the engineering solution of the bridge an »immense power« that could offer as much aesthetic pleasure as the conventionally beautiful object. »It was«, he said, »a tremendous problem in science to span the River Forth ... and the form the bridge took was probably as necessary for the solution of that problem as the form of the Pons Asinorum of Euclid. I cannot, therefore, at all sympathise with those who anathematise that wonderful structure for its ugliness. The effect it produces on my mind is a sensation of immense power, of a splendid human triumph over materials, which gives me as much pleasure as does a sensation of beauty.«[19] In drawing the distinction between our differing emotional responses to power and to beauty, Hodgson was pointing – wittingly or not – to the aesthetics of the sublime. And it is within the terms of this aesthetic category that it is possible to reconcile the deep emotional delight drawn by Anderson, Hodgson and their contemporaries from the scale and the powerful lines of the bridge. This reading is supported by Benjamin's Baker's own response to the view from the top of the three cantilever columns, which he described as »sublime«.[20]

In eighteenth-century Britain the beginning of the imperial adventure and the growth of industrialisation was marked by a powerful shift in aesthetic sensibility. Spurning the neo-classical delights of regularity, control and reason, writers like Lord Shaftesbury, Joseph Addison, and Edmund Burke found a new emotional and aesthetic satisfaction in the wild, the intense and the overwhelming. As Addison insisted in his best-known text on the matter: »Our imagination loves to be filled by an object, or to grasp at anything that is too big for its capacity. We are flung into a pleasing astonishment at such unbounded views, and feel a delightful stillness and amazement in the soul at the apprehension of them.«[21] In the mid-eighteenth century, Edmund Burke pointed to such qualities as infinity, vastness, power, difficulty, magnitude in building, and magnificence as stimuli to sublime contemplation. He also added the emotion of terror, or a potential source of terror viewed from a place of safety: »When danger and pain press too nearly, they are incapable of any delight, and are simply terrible; but at certain distances, and with certain modifications, they may be, and they are delightful, as we every day experience.«[22] The journey across the Forth Bridge, especially with the fate of the Tay Bridge lurking in one's mind, offered the late Victorians an exemplary display of the workings of the Burkean sublime: the immeasurable powers of nature challenged by a vast and powerful structure of a complexity that stretched beyond normal comprehension. The result, in Burke's words, could only be »astonishment ... admiration, reverence and respect«.[23]

Developing Burke's insights further, Kant identified two specific forms of sublime experience in his *Critique of Judgement*: the mathematical sublime, which is stimulated by extremes of scale or vastness; and the dynamic sublime, which results from the contemplation of displays of terrifying natural or physical power. Again, the Forth Bridge offers the perfect exposition of Kant's insights: a structure of vast dimensions designed to carry some of the most powerful machines ever constructed by man.

Although it is easy to list dimensions and statistics, the scale of the Forth Bridge defies easy comprehension since its visual impact changes constantly. As one moves around and across the bridge it assumes radically different appearances as its simple profiles are challenged by complex interior sections. In a paper given to the British Association in 1882, Benjamin Baker attempted to give an impression of its relative scale by comparing it with the longest bridge of the previous generation, the Britannia Railway Bridge built across the Menai Straits by George Stephenson in 1850. Discovering a ratio of 1 to 3.65 between the two structures, Baker hit upon the happy comparison between the average length of a new-born baby (19.33 inches in the 1880s, according to the Anthropometric Committee) and the height of the guardsmen recently sent to Egypt (5 feet 10.5 inches), giving the same ratio of 1 to 3.65. As Baker concluded: »As a Grenadier Guardsman is to a new-born infant, so is the Forth Bridge to the largest railway bridge yet built in this country.«[24] Baker's rule of thumb, however, is of little help in establishing the absolute dimensions of the bridge. A more useful aid is the drawing that superimposes the bridge on C. R. Cockerell's painting *The Professor's Dream* (1848), thus comparing it with the Great Pyramid at Giza, St. Peter's in Rome, St. Paul's Cathedral in London, the Pantheon in Rome, the Leaning Tower of Pisa, and the Central Transept of the Crystal Palace. By happy coincidence, Kant offered the experience of St. Peter's as an example of the mathematical sublime, suggesting that even

finite objects can produce sublime responses in the observer. According to Kant, when a visitor enters St. Peter's in Rome, »a feeling comes home to him of the inadequacy of his imagination for presenting the idea of a whole within which that imagination attains its maximum, and, in its fruitless efforts to extend this limit, recoils upon itself, but in doing so succumbs to an emotional delight«.[25] This harmonious tension between what is perceptually overwhelming and what is nevertheless known to be artifice provides the basis for a »specifically artistic sense of the sublime«.[26]

While the sublimity of the bridge exists in Kant's mathematical category in its sheer dimensions, it also draws on the dynamic sublime as soon as its function is brought into consideration. As Kant makes clear, the sublime exists not in the observed object itself but in the response of the observer. Nevertheless, certain phenomena are more likely to provoke strong reactions than others. In the context of the dynamic sublime, the individual confronts powerful and terrifying natural forces, and Kant himself offers a list of likely candidates: »Bold, overhanging, and, as it were, threatening rocks, thunderclouds piled up the vault of heaven, borne along with flashes and peals, volcanoes in all their violence of destruction, hurricanes leaving desolation in their track, the boundless ocean rising with rebellious force, the high waterfall of some mighty river, and the like, make our power of resistance of trifling moment in comparison with their might. But provided our position is secure, their aspect is all the more attractive for its fearfulness; and we readily call these objects sublime, because they raise the forces of the soul above the heights of the vulgar commonplace, and discover within us a power of resistance of quite another kind, which gives us courage to measure ourselves against the seeming omnipotence of nature.«[27] As the nineteenth century progressed, industrial production, the speed and power of steam technology, and the burgeoning metropolis or industrial city stimulated the sensations of awe, terror and exaltation previously associated with such natural phenomena

Notes

[1] The essential bibliography of the Forth Bridge would include: Benjamin Baker, *The Forth Bridge*, Spottiswoode, London, 1882; Wilhelm Westhofen, *The Forth Bridge*, Engineering Magazine, London, 1890, reprint: Moubray House, Edinburgh, 1989; *Industries: Forth Bridge Special Number*, 3 March 1890; Philip Phillips, *The Forth Bridge in its Various Stages of Construction and Compared with the Most Notable Bridges of the World*, R. Grant, Edinburgh, 1893; G. Backhausen, *Die Forth-Brücke*, Julius Springer, Berlin, 1889; Rolt Hammond, *The Forth Bridge and its Builders*, Eyre and Spottiswoode, London, 1964; Anthony Murray, *The Forth Railway Bridge: A Celebration*, Mainstream, Edinburgh, 1983; Sheila Mackay, *The Forth Bridge: A Picture History*, Moubray House, Edinburgh, 1990, revised edition: HMSO, Edinburgh, 1993; Roland Paxton (ed.), *100 Years of the Forth Bridge*, Thomas Telford, London, 1990; Arnold Koerte, *Two Railway Bridges of an Era: Firth of Forth and Firth of Tay: Technical Progress, Disaster and New Beginning in Victorian Engineering*, Birkhäuser, Basel and Boston, 1991; Charles McKean, *Battle for the North: The Tay and Forth Railway Bridges and the 19th-Century Railway Wars*, Granta, London, 2006.
[2] See: *Parliamentary Papers: Report of the Commission Appointed to Inquire into the Application of Iron to Railway Structures*, London, 1849.
[3] William McGonagall, *Poetic Gems*, 1890, reprint: David Winter, Dundee, 1973, pp. 39, 42.
[4] Theodor Fontane, *Sämtliche Werke*, vol. 20, Nymphenburger, Munich, 1962, p. 165. Fontane reported that his poem caused a sensation in Berlin, »perhaps greater than anything that I have written«. See: Helen Chambers, »Fontane's translation of ›The Charge of the Light Brigade‹, in: R. Byrn and K. Knight (eds.) *Anglo-German Studies*, Leeds, 1992, p. 96.
[5] Quoted from: Arnold Koerte, op. cit., (note 1), p. 108.
[6] John Eric Erichsen, *On Railway and Other Injuries of the Nervous System*, London, 1866, p. 9. Quoted from: Wolfgang Schivelbusch, *The Railway Journey: The Industrialization of Time and Space in the 19th Century*, Berg, Leamington Spa, 1986, pp. 141–142.
[7] Lewis Carroll, *Through the Looking Glass*, Penguin, Harmondsworth, 1994, p. 38.
[8] The Forth Bridge Railway Company was an independent undertaking, supported by the four leading Railway Companies: Midland, Great Northern, North Eastern, and North British.
[9] Wilhelm Westhofen, op. cit. (note 1), p. 1.
[10] Alfred Waterhouse, letter to John Fowler, quoted from: Sheila Mackay, op. cit. (note 1), pp. 111–112.
[11] *The Times*, 5 March 1890.
[12] *Industries: Forth Bridge Special*, loc. cit. (note 1), p. 52.
[13] *The Times*, 4 March 1890.
[14] Immanuel Kant, *Kritik der Urteilskraft*, 1790, 6th edition: Felix Meiner, Leipzig, 1924, p. 77.
[15] R. Rowand Anderson, in: *Transactions of the National Association for the Advancement of Art and its Application to Industry: Edinburgh Meeting 1889*, National Association ..., London, 1890, p. 153.
[16] William Morris, *News from Nowhere and Selected Writings and Designs*, Penguin, Harmondsworth, 1984, p. 332.
[17] Ibid., p. 188.
[18] Quoted from: Michael Baxandall, *Patterns of Intention*, Yale UP, New Haven, 1985, pp. 24, 25.

19 J. E. Hodgson, in: *Transactions of the National Association for the Advancement of Art and its Application on Industry: Birmingham Meeting 1890*, London, National Association ..., 1891, p. 11.

20 Sheila Mackay, op. cit. (note 1), p. 73.

21 Joseph Addison, *The Spectator*, no. 412 (23 June 1712).

22 Edmund Burke, *A Philosophical Enquiry into the Origin of our Ideas of the Sublime and Beautiful*, 1757, reprint: Routledge and Kegan Paul, London, 1958, pp. 39–40.

23 Ibid., p. 57.

24 Benjamin Baker, paper read before the British Association, Southampton, 1882; quoted from: *Industries ...*, op. cit. (note 1), p. 7

25 Immanuel Kant, *The Critique of Judgement*, trans. by J. C. Meredith, Oxford University Press, Oxford, 1973, p. 100.

26 See: Paul Crowther, *The Kantian Sublime*, Clarendon Press, Oxford, 1989, pp. 153–154.

27 Immanuel Kant, *The Critique of Judgement*, op. cit. (note 25), pp. 109, 110.

28 Wilhelm Westhofen, op. cit. (note 1), p. 65.

29 Théophile Gautier, *Histoire de Romantisme*, 3rd ed. Paris, 1877, p. 371, quoted from: John Gage, *Turner: Rain, Steam and Speed*, Allen Lane, London, 1972, p. 33. Gautier's demonic image recalls Milton's description in *Paradise Lost* of Satan's bridge-building skills: »Sin and death amain / Following his track, such was the will of heaven / Paved after him a broad and beaten way / Over the dark abyss, whose boiling Gulf / Tamely endured a bridge of wondrous length / From hell continued reaching the utmost orb / Of this frail world; by which the spirits perverse / With easy intercourse pass to and fro / To tempt or punish mortals, except whom / God and good angels guard by special grace.« John Milton, *Paradise Lost*, Book 2, ed. by Alastair Fowler, Longman, London, 1984, p. 138.

30 Friedrich Theodor Vischer, *Über das Erhabene und Komische und andere Texte zur Ästhetik*, Suhrkamp, Frankfurt am Main, 1967, p. 155.

The reproductions of all Evelyn Carey photographs by kind permission of the British Railways Board.

6. Cockerell, the Forth Bridge compared with other celebrated buildings. (Civil Engineering Department Library, Imperial College, London.)

7. J. M. W. Turner, *Rain, Steam and Speed*, 1844. (National Gallery, London.)

as cliffs, waterfalls and deserts. As the branch of high technology most accessible to the wider public, railway travel featured highly on the list of sensations guaranteed to provoke sublime sentiments by offering us, in Kant's words, the chance to measure ourselves against nature. Following the opening in 1830 of the Liverpool and Manchester line, the world's first commercial railway, Britain became gripped by railway mania in the 1840s. In 1846 alone Parliament authorized the railway companies to raise £ 132,000,000, representing only a quarter of the total number of schemes, wild to various degrees, that had been submitted to the Board of Trade. In his text on the Forth Bridge, Westhofen pointed to the railway mania of the 1840s as a formative influence on the young John Fowler, describing »how fortunes were made and ruined in a day; how men lost their reason in a moment both from good and evil tidings, and how the capital subscribed during those years, often for the wildest undertakings, almost rivalled the days of the South Sea Bubble«.[28] In 1850 there were 6,635 miles of track in Britain, increasing to 15,319 miles in 1870, and to 20,073 miles in 1890. As an extreme example of the dynamic and technological sublime, the railway mania undermined all notion of restraint or limitation and proposed, instead, a boundless perspective of continual innovation and transformation. The overwhelming commercial, geopolitical, and industrial might of nineteenth-century Britain was most powerfully embodied in the railways, whose snorting and dashing locomotives and athletic bridges and viaducts annihilated space and time. No longer was the roar of the waterfall or the peal of thunder the necessary accompaniment to ravishing sublime insight, but rather the shriek of escaping steam and the rumble of iron wheels.

The analogy between the natural and the technological sublime is the subject of one of J. M. W. Turner's most celebrated canvases, *Rain, Steam and Speed*, painted in 1844. It portrays a steam train hurtling at full speed across a storm-wracked bridge, high above the heaving water of the river below. The French writer and critic Théophile Gautier described it brilliantly as »a real cataclysm. Flashes of lightning, wings like great fire-birds, towering columns of cloud collapsing under the thunderbolts, rain whipped into vapours by the wind. You would have said it was the setting for the end of the world. Through all this writhed the engine, like the Beast of the Apocalypse, opening its red glass eyes in the shadows, and dragging after it, in a huge tail, its vertebrae of carriages.«[29] While early confrontation with the railways stimulated fear, terror, incomprehension and nervous collapse, the overcoming of these responses was accompanied by a vigorous self-assurance as the railway decades progressed. Writing in the mid-nineteenth century, the German aesthetician Theodor Vischer stressed the positive and constructive aspect of the sublime in terms that perfectly reflect the triumph of the engineers: »We feel ourselves elevated because we identify ourselves with the powers of nature, ascribing their vast impact to ourselves, because our fantasy rests on the wings of the storm as we roar into the heights and wander into the depths of infinity. Thus we ourselves expand into a boundless natural power.«[30] Yet the fascination of terror and danger lingered. At 3:30 am on the very day that the Prince of Wales opened the Forth Bridge, the London to Glasgow express crashed outside Carlisle station. Four passengers died, one of whom, as *The Times* was keen to point out the following day, was a lady who »had her throat dreadfully cut«. Little wonder that early passengers crossing the Forth Bridge, while revelling in the magnificence of the structure, threw pennies out of the carriage window for good luck.

Angus J. Macdonald

The technology of the Forth Bridge

The designers of the Forth Bridge were presented with two formidable problems. One was the very long spans which were required to cross the two deep channels of the river. The other was the need to avoid blocking these channels, which were required for navigation while the bridge was under construction. Both problems were brilliantly solved by John Fowler and Benjamin Baker.

Including its approach viaducts, the Forth Bridge is 1 mile, 1,005 yards long, while the cantilever portion from pier to pier totals 5,300 feet, which is 944 feet more than the height of Scotland's premier mountain, Ben Nevis. Exceeding the hitherto longest span in the world – New York's Brooklyn Bridge – by 114.5 feet, the two great spans are each 1,710 feet. At the point of the bridge crossing the cross-section of the river valley consists of two deep river channels, each some 600 feet wide, separated by the rocky island of Inchgarvie. To the south there is an area of shallow water and a narrow strip of land at sea level, while to the north there is only a narrow strip of low-level land. On both north and south shores the narrow coastal strip is bounded by a steep bank where the land rises to the approach level of approximately 160 feet above high water. The height of these approaches, together with the navigational requirements of the river, dictated a rail crossing at high level, and the final design gave a rail level height of 157 feet above the high water mark.

The bridge is in three parts: short-span viaducts were used at the north and south shores over the low coastal strips where the building of foundations was relatively simple. A cantilever-and-suspended-span configuration was adopted for the long spans across the deep river channels. This consisted of three pairs of balanced cantilevers and two suspended spans. The balanced-cantilever arrangement, the basic form of the bridge illustrated in ill. 2, was adopted so that the bridge would be self-supporting throughout the sequence of construction. A simple beam structure built *in situ* would have demanded temporary supports during construction. Alternatively, a beam structure could have been built on the shore near the bridge, floated out to the site on pontoons and then hoisted into its final position on its foundation piers. The latter method was frequently adopted for the erection of large girder bridges in the nineteenth century: both the Royal Albert Bridge at Saltash and the Britannia Bridge across the Menai Straights were constructed in this way. Neither method was feasible at the Forth Bridge, however, due to a combination of the bridge's immense size and to the need to keep the navigation channels free of obstruction during the construction process. The use of balanced-cantilever construction allowed the bridge to be built *in situ* without the need for temporary structures.

The concept

It is a truism of structural engineering that the longer the span the higher must be the level of efficiency of the structure. This is because, for a horizontally spanning structure of a given configuration carrying a gravi-

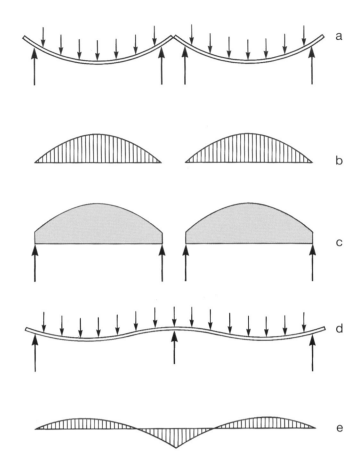

tational load, the ratio of the self weight to the load carried tends to increase as the span increases.[1] Given the very long spans of the principal parts of the Forth Bridge, it was therefore essential that a high level of structural efficiency be achieved.

For the basic configuration of the bridge Fowler and Baker favoured a horizontally spanning girder forming pairs of balanced cantilevers. This structural form, however, is not particularly efficient and was chosen to satisfy the need for a method of construction that would not block the navigational channels, rather than from a consideration of the efficiency of its structural performance. For in terms of efficiency, arches or cable suspension systems are fundamentally more efficient. Given the long spans involved, a design strategy was therefore necessary which would ensure that an appropriate level of efficiency was achieved. These were of a highly innovatory nature and were responsible, as we shall see, for the importance of the bridge in the history of engineering. The three most significant innovations were the use of the continuous-girder principle, the matching of the elevational profile of the structure to the distribution of the internal forces generated by the gravitational load, and the full triangulation of the internal geometry of the structure.

The continuous-girder principle

Although the structural geometry of the Forth Bridge looks complex in the foreshortened view from the shore, its side elevation is remarkable both for its simplicity and for the clarity with which the engineering principles on which the bridge was based are expressed. This is the result of the very clear theoretical understanding of structural behaviour possessed by the principal designer Benjamin Baker, who configured the bridge as a statically determinate continuous girder. The structural continuity allowed high efficiency to be achieved in the use of material (a necessity for a long span structure), while the statical determinacy eliminated the problems of »lack of fit« and of stress due to thermal expansion, which are normally associated with continuous structures.[2] The advantages of the continuous girder over the simply supported girder were known to mathematicians working in the field of structural theory by the beginning of the nineteenth century, but their ideas were only slowly adopted into practice. Baker himself had written a series of articles in the 1860s on the application of this theory to the design of bridges, and the Forth Bridge was one of the first major structures in which this new knowledge was put into practice.

The essential features of the continuous girder can be understood by examination of very simple beam-type structures. The effect of gravitational load on a beam is to bend it. The ends of the structure rotate about their supports and remain straight, though tilted, while the parts towards mid-span become curved (ill. 1). The intensity of curvature varies from a maximum at the mid-span point to a minimum at the supports. The strength of cross-section required to resist the bending is directly proportional to the intensity of curvature. A diagram showing the distribution of this intensity across the span can be determined from the load pattern, and is called the bending-moment diagram, as bending moment is the term used to de-

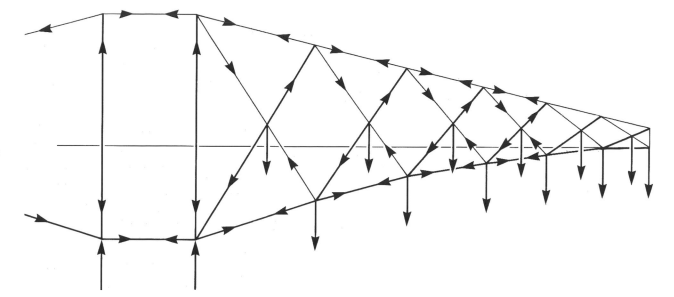

1. In a bridge made from two simply supported beams there is a discontinuity at the central support (a). The distribution of the internal forces which occur (bending moments) is shown in (b) while (c) shows the ideal profiles for the bridge girders in which the depth is varied to correspond with the diagrams of bending moment. This results in material and therefore dead weight being concentrated at the two span positions. Where the two-span bridge is continuous over the central support (d) a reversal of curvature occurs in the deflected form. This results in a corresponding reversal in the sense of the internal forces (bending moments), and favours a different and more efficient distribution of dead weight than with simply supported beams. A further advantage is that the maximum values of the internal forces, for a given intensity of load, are smaller.

2. A balanced cantilever arrangement is obtained by making girders continuous over the supports and inserting hinges at the mid span points (a). This results in a concentration of internal force over the supports and reduces the mid span bending moments to zero (b). If two hinges are inserted between each support a cantilever and suspended span arrangement is produced (c). This results in a more even distribution of internal force than occurs with balanced cantilevers (d). This is the basic configuration of the main spans of the Forth Bridge. In the Forth Bridge (e) the overall profile is matched to the distribution of internal force (bending moment) of the cantilever and suspended span carrying a uniformly distributed load. This results in the material of the bridge being concentrated at the locations where the internal forces are highest and ensures that loads are carried efficiently.

3. The internal geometry of the main girders of the Forth Bridge is fully triangulated. The loads from the internal viaduct are introduced only at the vertices of the triangles which ensures that the internal forces in the sub-elements are axial.

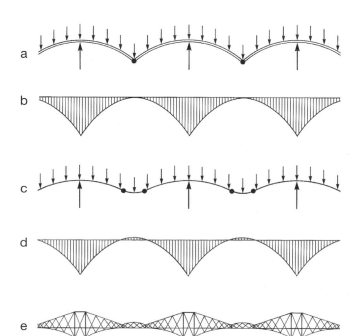

a
b
c
d
e

scribe the internal force which is generated by the curvature of an initially straight beam.

Where high efficiency is required the designer must ensure that strength is provided only where it is needed. In a simply supported beam bridge the bending moment is greatest at mid span, so it is here that most of the material is concentrated in the form of a larger cross-section and greater structural depth. This makes the bridge heavier at mid span than over the supports, an undesirable result since the material of the bridge forms part of the load which it must carry, and loads applied at mid span have a greater tendency to cause bending than those applied near the supports.

If a girder is made continuous across several supports, hogging curvature occurs at the supports and reduces the degree of sagging curvature in the middle of the spans (ill. 1, d). More strength is required at the supports, but less at mid span. At the points where the direction of curvature changes – the points of contra-flexure – the girder remains straight, though tilted, and at these points no bending strength at all is required. Exploiting these characteristics, the continuous girder makes possible a structure in which the curved parts of the girder, where the strength is required, are at the supports, while the straight parts are not at the points of support. This allows a more satisfactory distribution of self weight to be achieved than is possible with a simple beam.

The continuous girder offers other advantages. Compared with a series of simply supported beams, the continuous girder reduces the maximum values of the internal forces (bending moments) for a given intensity of load. Although the total depth of the bending-moment diagram is the same as for the simply supported beam – being determined by the span and the intensity of the load – the actual peak values of bending moment in the continuous girder are less because the reversal of curvature has the effect of pulling the bending-moment diagram under the zero line of the diagram where the reversal of curvature occurs (ill. 1, e). This means that less strength, and therefore less structural material, need be provided for a given load-carrying capacity to be achieved.

There are, however, also disadvantages. Small changes in the support conditions such as might occur due to differential settlement of foundations or as

a result of thermal expansion of the structure cause an increase in internal force to occur. This problem, however, can be countered by incorporating into the structure discontinuities in the form of hinge-type connections at the points where contra-flexure occurs. A hinge in fact acts as a point of contra-flexure so hinges can be used to locate points of contra-flexure at the most convenient places across the span. In this way, the shape of the bending-moment diagram and thus the most desirable profile for the structure can be manipulated by varying the positions where such hinges are included.

If the hinges are set at the mid-span position of a continuous girder a balanced-cantilever configuration is produced, in which the maximum curvature under load occurs over the supports and the minimum curvature occurs at mid span (ill. 2). The maximum strength and self weight are therefore concentrated at the supports where the weight can be most easily resisted.

In the main spans of the Forth Bridge two hinges were inserted between each pair of supports to form the cantilever-and-suspended-span arrangement. Although this introduced some bending into the mid span position it reduced the total self weight of the structure because locating the hinges nearer to the supports reduced the amount of curvature (bending) at the supports, as can be seen from the bending-moment diagram (ill. 2, d).

Triangulation of the internal geometry

In the mid-nineteenth century, techniques for analysing triangulated geometries were developed by engineers and mathematicians such as Squire Whipple, Carl Culmann, August Ritter and James Clark-Maxwell. That Baker understood this research is amply demonstrated by the clear way in which the internal geometry of the Forth Bridge was triangulated with no redundant elements, no requirements for continuity through joints to provide stability, and the introduction of the loads to the main structure only at the vertices of the triangles.

The main elements of the Forth Bridge were built up from a series of smaller sub-elements which were arranged into a triangulated geometry (ill. 3). The purpose of this was to eliminate bending and replace it with a combination of axial tension and compression, since axial internal force can be resisted more efficiently than bending. Thus, although the structure of the Forth Bridge, taken as a whole, had a tendency to bend under the action of the load, the individual members are subjected either to axial tension or to axial compression. This is due to the geometric properties of the triangle, which is the only straight-sided polygon whose shape cannot be altered other than by changing the length of one or more of its sides. In contrast, the geometry of any other polygon may be changed without altering the length of any side by changing the angles between the sides.

If an attempt is made to alter the geometry of a triangle by subjecting it to load, the resistance to the load takes the form of a resistance to a length change of the sides, and the members forming the sides are therefore subjected either to direct tension (resisting a tendency to lengthen) or to direct compression (resisting a tendency to shorten). This only occurs if the load is applied to the vertices of the triangles. If it is applied directly to one of the sides then that side will be bent by its action.

If a structure is built up from triangles its resistance to deformation when a load is applied is similar to that of a single triangle. Thus, the efficiency of a beam-type structure can be improved by breaking it down into sub-elements that are arranged in a triangulated pattern. This is precisely what was done at the Forth Bridge: all the large girders in the structure were triangulated and the general arrangement was such that the external loads were introduced into the structure only at the vertices of the triangles.

Steel

Fowler and Baker's design depended not only on the most advanced techniques of structural design, but also on the most modern material of the day: steel. The Forth Bridge was the first major structure in Europe to be built of steel, a far superior material to the cast and wrought iron of earlier nineteenth century metal bridges.[3] It first became available as a structural material following the introduction of the Bessemer process in the mid-1850s and the development of the Siemens open hearth furnace in the late 1860s. Compared with cast or wrought iron, steel had much better strength properties and was available to the builder in the form of hot-rolled components, ranging in shape from flat plates of various thicknesses to linear elements with a wide range of cross-sectional shapes, such as the angle, channel, T and double-T sections. Similar ranges had been produced in wrought iron but a significant difference in the case of steel was that the sizes of individual elements could be much larger than had been possible with wrought iron. This difference was a consequence of the processes by which the materials were manufactured. Wrought iron could only be made in small batches, which limited the maximum size of components to a weight of around 100 pounds. It meant that large wrought-iron structures, such as the Britannia Bridge, had to be constructed by riveting together vast numbers of small components. With steel, individual components weighing several tons each could be manufactured, and the limiting factor on the maximum feasible size of a component became one of handling and transportation rather than of the manufacturing process.

4. The junctions between the sub-elements are of complex geometry. Although heavy machines such as multiple drills and plate bending machines were used, the bridge was essentially hand crafted. Photo: Eric de Maré. (Royal Commission on the Ancient and Historical Monuments of Scotland.)
5. The insertion of cross-bracing between the pairs of tubes which act as compression members produced combined struts which were capable of resisting lateral load and which became the wind bracing elements of the structure.
6. Wind loading on the bridge is conducted to the bases of the main towers by the braced inclined struts and lower booms of the cantilevers.
7. Living model illustrating the principle of the Forth Bridge. (Civil Engineering Department Library, Imperial College, London.)

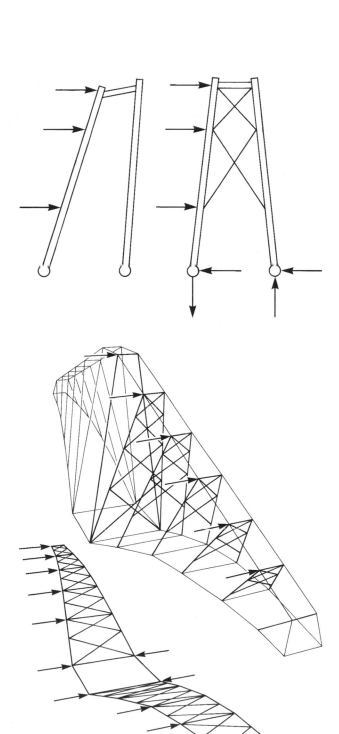

The adoption of steel was a contributory factor in the development of a relatively simple elevational geometry for the Forth Bridge. The great overall size of the structure meant that a simple configuration with a small number of relatively large sub-elements carrying large amounts of load could only be adopted if a very strong material was employed. The form of the Forth Bridge was, therefore, as much a tribute to the properties of steel as it was to the structural principles on which its design was based.

Construction

Although the geometry of the elevation of the Forth Bridge is relatively simple, the overall form of the bridge is in fact fairly complex because the cross-sectional and plan shapes of the principal girders are tapered. This means that the geometries of the junctions between the sub-elements are very complicated (ill. 4) and, for this reason, the construction of the bridge was potentially very difficult. Fowler and Baker were fortunate in having appointed as chief contractor to the bridge William Arrol, a remarkable individual whose efficiency and innovative solutions to the problems posed by the construction of the bridge contributed in no small measure to its success.

Before its final erection above the Forth, the superstructure of the bridge was first fabricated in sections in a very large erecting yard on the south shore. The individual plates and linear elements were first shaped and drilled and then bolted together to form the various sub-assemblies of the bridge. This allowed the viability of the various complex joints linking these assemblies to be checked. These major components were then dissembled and re-erected in their final positions in the bridge superstructure. The preliminary reconstruction was carried out by bolting, but each erecting gang was followed by a team of riveters who replaced the bolts one by one with the final riveted fastenings, some 7,000,000 in all.

Stability and bracing

When a load is exerted on the central girder of the Forth Bridge the upper booms of the cantilevers come into tension, while the lower booms are compressed. Benjamin Baker gave a famously graphic demonstration of this principle at a lecture given in 1889 at the Royal Institution in London, using two men, two chairs, two piles of bricks, four broomsticks, and a Japanese engineer named Kaichi Watanabe (ill. 7). The relationship of stretching and compression was given powerful articulation in Baker's design in the use of lattice girders composed on L-section beams for the upper booms and oval or circular tubes for the lower booms: the best cross-sectional shapes for resisting compression.

The susceptibility of compression elements to buckling (the principal structural problem which has to be overcome with this type of element) depends on their length, or, more exactly, on their slenderness (the ratio of their length to their width). In the Forth Bridge, the slenderness ratio of individual tubes was reduced by connecting pairs of tubes together with cross-bracing to form combined struts (ill. 5). This was done with both the inclined internal members and in the lower booms of the cantilevers.

The combined struts are capable of resisting lateral as well as axial load and so the braced inclined internal members, together with the braced lower booms of the cantilevers are used to resist wind loading acting on the side of the bridge (ill. 6). The absence of any bracing in the top booms (tension members have no tendency to buckle and do not need to be braced) rendered them incapable of resisting any horizontal load. This allows the top of the bridge to sway in response to the wind and ensures that all wind loads are transmitted directly to the bases of the main towers where they can be most effectively resisted. The ingenious configuration of the cross-bracing is one of the most satisfying aspects of the design of the bridge.

Expansion and contraction

A large metal structure like the Forth Bridge undergoes a considerable amount of expansion and contraction in response to changes in temperature. The 1710-foot span, for example, expands a distance of 2.64 inches between the months of January, when the mean temperature is + 4° C and July, when the mean temperature is +15° C. This would be sufficient to cause a stress of 1.6 ton/sq inch in the steelwork in addition to the stress already there due to the applied loads. Such extreme temperature differences of + 25° C and –10° C demand an expansion of 8.4 inches over the seasons, resulting in a stress of 5 ton/sq inch. To prevent this additional »temperature« stress from occurring, the bridge contains expansion joints that allow the lengths of the spans to vary. There are four expansion joints in the superstructure. Two occur at the extreme ends of the cantilevered part of the structure, where the cantilevers enter the stone approach portals, and two are positioned at the junctions between the central cantilever girder and the suspended spans at either end of it (ill. 11).

The design of the expansion joints is of considerable interest. It will be remembered that the joints between the cantilevers and the suspended spans must behave as hinges in order that the cantilever part of the bridge, taken as a whole, will behave as a statically determinate continuous girder. The joints must therefore be capable of transmitting horizontal and vertical forces but not bending. They must also allow com-

plete freedom for longitudinal expansion to occur. An ingenious design, in which the suspended spans were mounted on rocking posts fitted into recesses in the ends of the cantilevers, was worked out for the joints between the suspended spans and the cantilevers (ills. 8). The rocking posts transmit the vertical load between the suspended span and the cantilever without restricting longitudinal movement. Lateral loads are transmitted through vertical pins which are attached to the cross girders (top and bottom) at the ends of the suspended spans and mounted in square blocks which slide in brackets fixed to the cantilevers. These too allow complete freedom for longitudinal expansion.

The expansion joints at the points where the ends of the cantilevers enter the stone approach portals are simpler than those at the suspended spans. Here the ends of the cantilevers simply rest on roller bearings that allow longitudinal movement. Horizontal forces are transmitted by direct-contact sliding bearings.

As a further precaution against damage due to thermal expansion, provision was made for longitudinal movement to occur at the interface between the bridge superstructure and the solid masonry piers on which it rests. This prevents temperature stresses from occurring in the horizontal members at the bases of the towers which form the central parts of each pair of balanced cantilevers. Each foundation junction consists of a lower bedplate, bolted to the masonry foundation, and an upper bedplate riveted to the base of the superstructure of the bridge. Key plates were

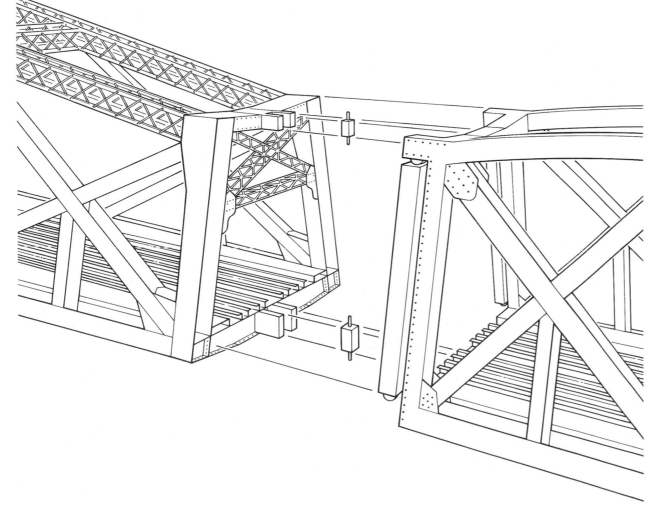

8. Expansion joint at the junction of the suspended spans and the Inchgarvie cantilever. The weight of the suspended span is carried by rocking posts which are located in hollow tubes at the ends of the cantilever. Pins, attached to the suspended span, are held between pairs of blocks which protrude from the cantilever. The joint allows movement in the longitudinal direction while transmitting both lateral and vertical force. The junction behaves as a hinge, however, because rotation of the main elements relative to each other is not restrained.
9. Junction between cantilever tower and foundation.
10. Three different configurations were used at the junctions between the cantilever towers and the foundations.
11. Locations of the expansion joints.
12. Distribution of foundation types at the bases of the cantilever towers.

Notes

[1] See: Angus J. Macdonald, *Structure and Architecture*, Butterworth, Oxford, 1994, chapter 6.

[2] Ibid., appendix 3 for an explanation of the problems of »lack of fit« and thermal expansion in relation to continuous structures.

[3] It was in fact the second major structure in the world in which steel was used – the first being the three-arch bridge over the Mississippi at St. Louis by James Eads, completed in 1874.

[4] See: Eda Kranakis, *Constructing a Bridge: An Exploration of Engineering Culture, Design and Research in Nineteenth-Century France and America*, MIT Press, Cambridge, MA, 1997.

[5] See: William Pole (ed.), *The Life of Sir William Fairbairn*, 1877, reprint: David and Charles, Newton Abbot, 1970.

Explanatory line drawings by Stephen Gibson.

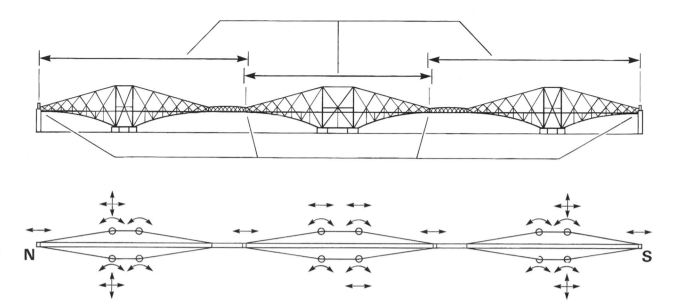

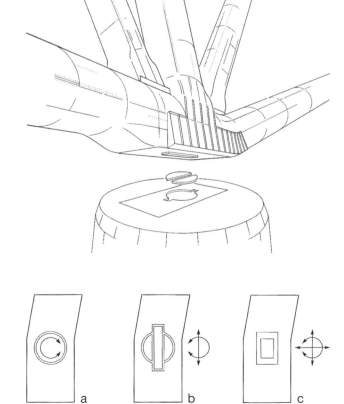

riveted to the upper bedplates and were accommodated in recesses in the lower bedplates (ill. 9). The precise shape of the key plates and their corresponding recesses determined the amount of movement which was possible at each foundation.

There are the following three types of foundation interface:
– a circular key plate riveted to the upper bedplate and fitting exactly into a circular recess in the lower bedplate (ill. 10, a) – this permits rotation only at the interface;
– a rectangular key plate riveted to the upper bedplate and two segmental floating key plates (ill. 10, b) – the recess in the lower bedplate was shaped to allow rotation and transitional movement in one direction only;
– a rectangular key plate riveted to the upper bedplate and fitting into a larger rectangular recess in the lower bedplate (ill. 10, c) – this permits rotation and translation in any direction.

In each case the lower bedplate was bolted to the masonry pier by 48 bolts 2.5 inches diameter which were sunk 25 feet into the masonry and anchored to cast-iron anchor plates. The nuts, which were tightened down firmly onto the lower bedplates, have tubular shanks and hexagonal heads. The shanks were located in slots in the upper bedplates and are of such a length as to leave the hexagonal heads 0.063 inches clear of the top of the upper bedplates. The upper bedplates and the bridge superstructure are therefore free to move independently of the anchor bolts which hold the structure down only if it tends to lift. As the weight of the bridge is very great this occurs only in the most extreme wind conditions. The three types of foundation were arranged at the bases of the towers in such a way that the steel superstructure can expand and contract in response to changes in temperature while at the same time being adequately fixed to its supports (ill. 12).

Conclusion

The Forth Bridge can be seen both as a structure which is a great work of engineering in its own right and as a historical artefact of cardinal importance. While some aspects of its design pointed the way forward to the twentieth century, others were part of a long-standing tradition of engineering craftsmanship. Its principal designer was one of the best, most forward-looking engineers of the late nineteenth century who operated at the very cutting edge of engineering design practice; its constructor was one of the best contractors in metalwork of the day – someone who was rooted in a craft tradition. The bridge stood therefore on the cusp of traditional and modern bridge building practice and therefore on the threshold of the modern world.

At the Forth Bridge, Baker used his theoretical knowledge of structural behaviour to generate the form of the structure and by doing so he introduced to Britain a new technique for the design of major structures. The method had been pioneered in continental Europe, and particularly in France, where much of the theory of structures was developed in the nineteenth century under the influence of the École Polytechnique.[4] The most notable practitioner was Gustave Eiffel, whose eponymous tower was erected in 1889, and who was also a guest at the official opening of the Forth Bridge in March 1890. The method was uncommon in Britain, however, where an empirical tradition existed, in which innovative structural forms were developed through trial and error by testing models and parts of structures: the series of tests which were carried out by William Fairbairn in connection with the revolutionary design for the Britannia Bridge across the Menai Straights in Wales is a prominent example.[5] By allowing theoretical principles to contribute to the evolution of the form of the Forth Bridge, Baker was using, in the context of one of the most spectacular and well-publicised works of engineering of its day, an innovatory technique. This method would become the normal means by which structures were designed in the twentieth century. The Forth Bridge was therefore important in the evolution of modern techniques of structural design because the publicity which surrounded this huge undertaking drew attention to the coming methodology.

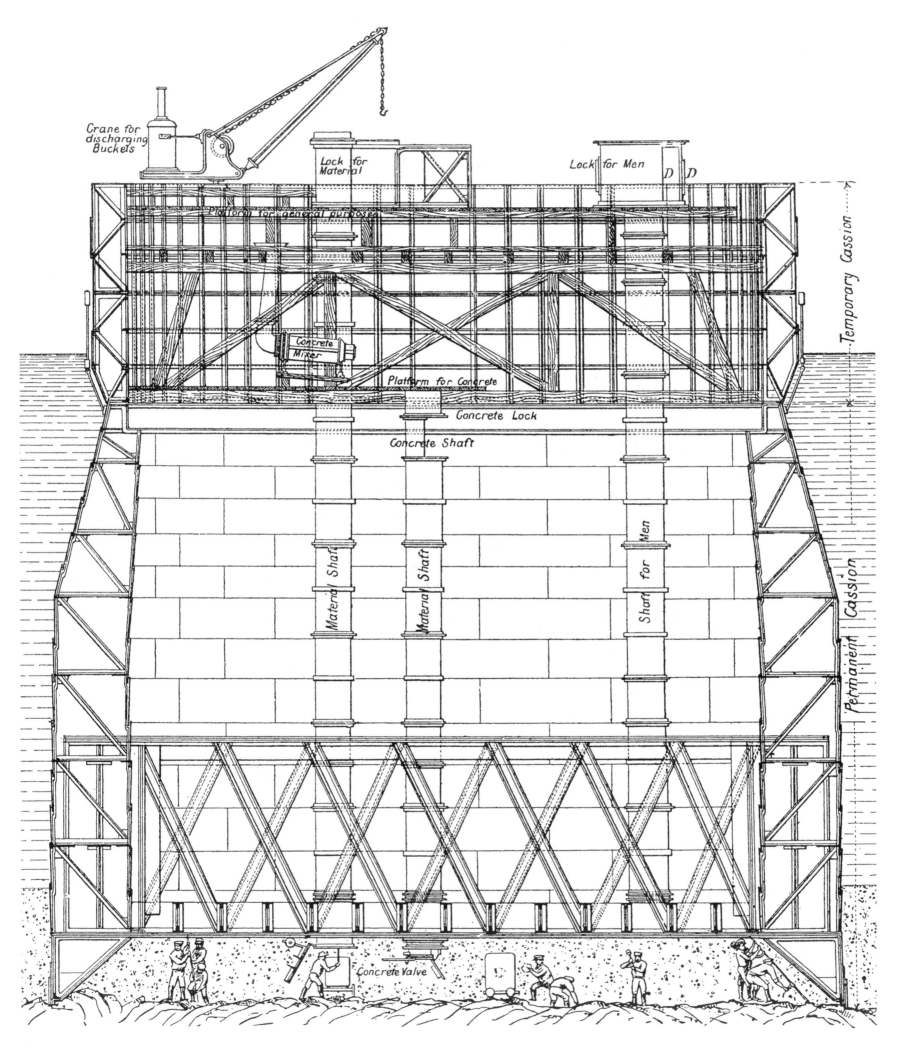

Crane for discharging Buckets

Lock for Material

Lock for Men

D D

Platform for general purposes

Temporary Caisson

Concrete Mixer

Platform for Concrete

Concrete Lock

Concrete Shaft

Material Shaft

Material Shaft

Shaft for Men

Permanent Caisson

Concrete Valve

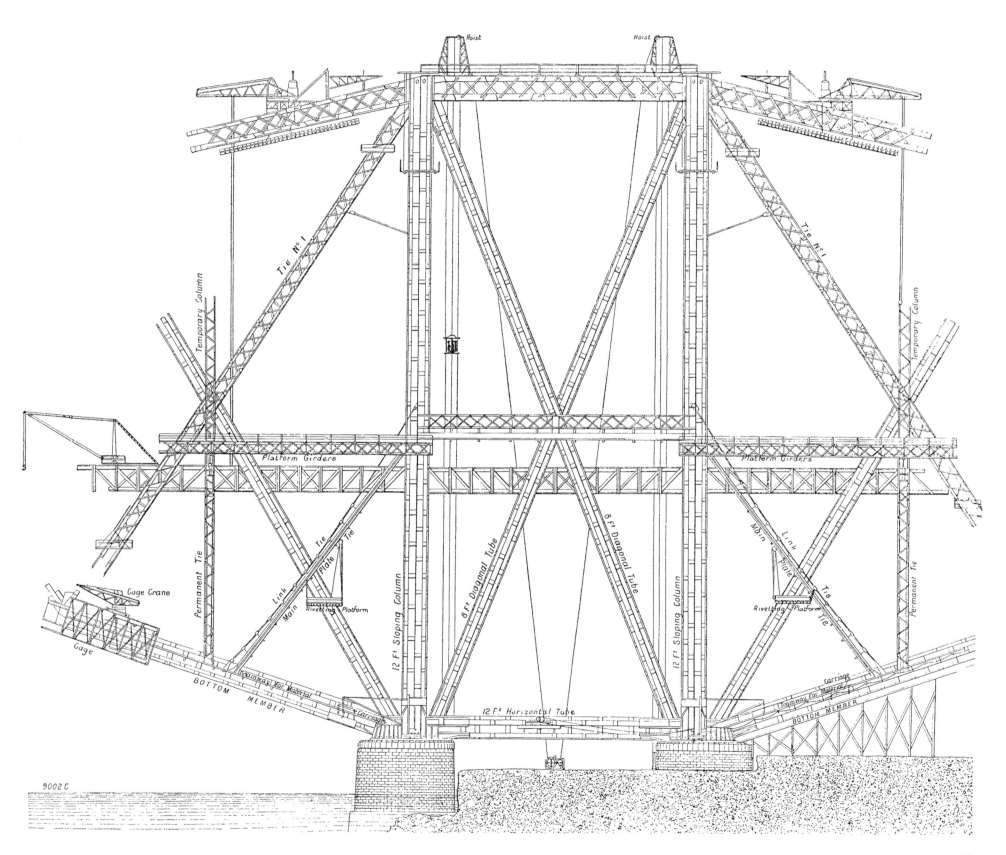

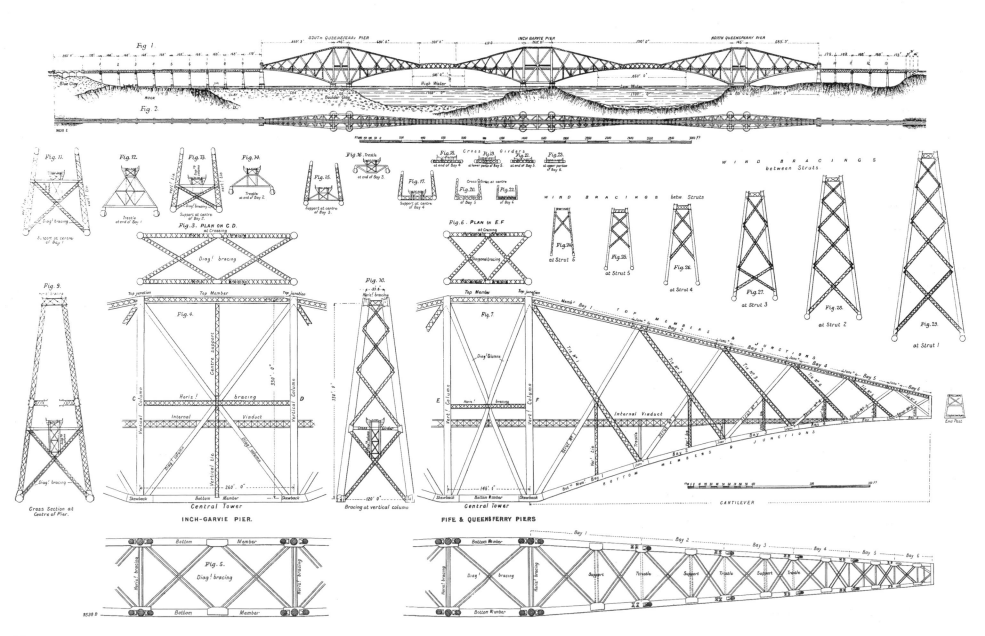

20

3. Elevation and plan with details of the steelwork.
(Wilhelm Westhofen, *The Forth Bridge*, 1890.)
4. Elevation and sections of the complex skewbacks
at the bases of the main towers. (Wilhelm Westhofen,
The Forth Bridge, 1890.)

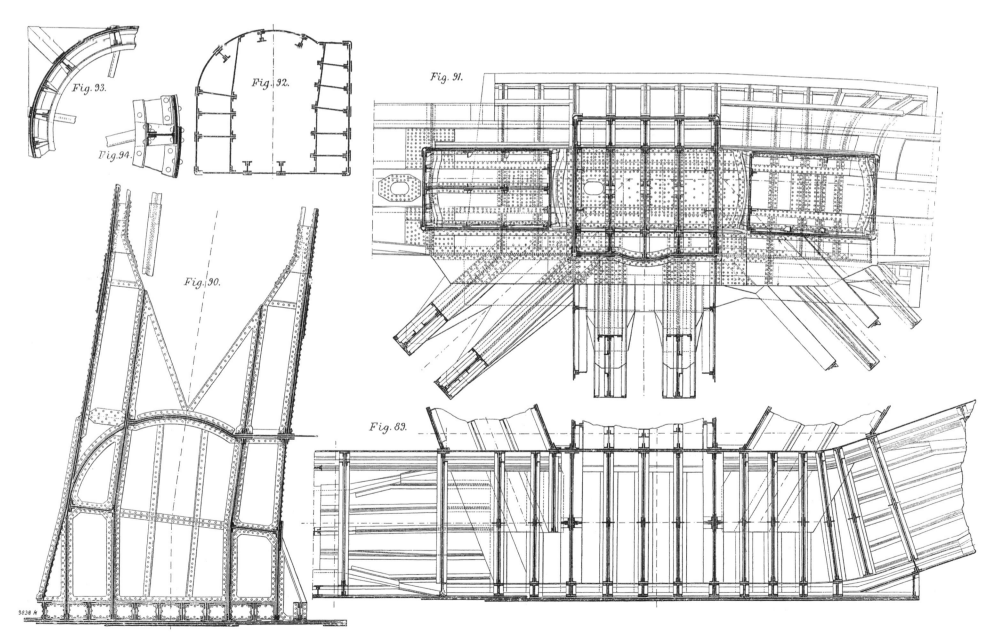

Statistics

Length
Cantilever bridge total: 5,300 feet
Base of Inchgarvie tower: 260 feet
Base of Fife and Queensferry Towers both: 145 feet
Suspended spans both: 350 feet
Six cantilevers each: 680 feet
South approach viaduct from centre of cantilever
    end pier: 1,978 feet
North approach viaduct: 968 feet 3.5 inches
Total length of structure: 8,295 feet 9.5 inches

Height above and depth below high water mark
Top of 12 granite piers: 18 feet
Rail level height unloaded: 157 feet
Top of central towers: 361 feet
Greatest depth to which foundations were
    carried: 88 feet

Breadth
Between base of central towers: 120 feet
Between tops of central towers: 33 feet
    (inclination of 1 in 7.5)

Materials
Total weight of steel in main spans: 50,958 tons
Portland cement: 21,000 tons
Aberdeen granite: 740,000 cubic feet (56,100 tons)
Rivets: ca. 7,000,000
Painted area, internal and external: 145 acres

Schedule
Contract let: 21 December 1882
Completion of South Jetty: May 1884
Launching of sixth and last caisson: May 1885
Completion of 12 main pier foundations: June 1886
Raising of viaduct piers and girders: July 1887
Start of erection of steel towers: Summer 1887
Completion of all 6 cantilevers: July 1889
Completion of suspended spans: November 1889
Test loading by two trains: January 1890
Official opening: 4 March 1890